# THE POETRY OF FRANCIUM

# The Poetry of Francium

Walter the Educator

Silent King Books

**SILENT KING BOOKS**

# SKB

Copyright © 2024 by Walter the Educator

All rights reserved. No part of this book may be reproduced in any manner whatsoever without written permission except in the case of brief quotations embodied in critical articles and reviews.

First Printing, 2024

Disclaimer
This book is a literary work; poems are not about specific persons, locations, situations, and/or circumstances unless mentioned in a historical context. This book is for entertainment and informational purposes only. The author and publisher offer this information without warranties expressed or implied. No matter the grounds, neither the author nor the publisher will be accountable for any losses, injuries, or other damages caused by the reader's use of this book. The use of this book acknowledges an understanding and acceptance of this disclaimer.

"Earning a degree in chemistry changed my life!"
– Walter the Educator

dedicated to all the chemistry lovers, like myself, across the world

# FRANCIUM

In the heart of the periodic table's maze,

# FRANCIUM

Where elements dance in their mysterious ways,

# FRANCIUM

Lies Francium, a treasure untamed,

# FRANCIUM

In its atomic dance, it's hardly framed.

# FRANCIUM

A shy and fleeting element it seems,

# FRANCIUM

Elusive like dreams in midnight gleams,

# FRANCIUM

But oh, the tales it yearns to tell,

# FRANCIUM

Of atomic symphonies in its ethereal shell.

# FRANCIUM

From the depths of Earth, it silently rises,

# FRANCIUM

In the embrace of radium, it disguises,

# FRANCIUM

Its presence felt in traces so rare,

# FRANCIUM

Yet its significance, none can compare.

# FRANCIUM

With atomic number eighty-seven it shines,

# FRANCIUM

In the realm of chemistry, it defines,

# FRANCIUM

A metal so soft, it melts with ease,

# FRANCIUM

In the warmth of hands, it finds release.

# FRANCIUM

But beware, for Francium holds a fiery touch,

# FRANCIUM

A reactivity that can seem too much,

# FRANCIUM

With a half-life fleeting, it swiftly decays,

# FRANCIUM

In a dance of particles, it quietly sways.

# FRANCIUM

In laboratories, it's a coveted prize,

# FRANCIUM

For scientists with curious eyes,

# FRANCIUM

To glimpse its nature, to understand,

# FRANCIUM

The secrets it holds in its atomic band.

# FRANCIUM

Yet in nature's vast and ancient scroll,

# FRANCIUM

Francium hides, a rare and noble goal,

# FRANCIUM

In minerals deep within the Earth's embrace,

# FRANCIUM

A whisper of its presence, a trace.

# FRANCIUM

Oh Francium, element of mystery,

# FRANCIUM

In your atomic dance, a symphony,

# FRANCIUM

You captivate with your fleeting grace,

# FRANCIUM

A reminder of nature's hidden face.

# FRANCIUM

Though elusive and rare, you leave your mark,

# FRANCIUM

In the annals of science, in the dark,

# FRANCIUM

A testament to the wonders unseen,

# FRANCIUM

In the realm of elements, where you've been.

# FRANCIUM

So let us marvel at Francium's lore,

# FRANCIUM

A tale of chemistry's endless explore,

# FRANCIUM

In its fleeting presence, we find delight,

# FRANCIUM

In the dance of atoms, both day and night.

# FRANCIUM

# ABOUT THE CREATOR

Walter the Educator is one of the pseudonyms for Walter Anderson. Formally educated in Chemistry, Business, and Education, he is an educator, an author, a diverse entrepreneur, and he is the son of a disabled war veteran. "Walter the Educator" shares his time between educating and creating. He holds interests and owns several creative projects that entertain, enlighten, enhance, and educate, hoping to inspire and motivate you.

Follow, find new works, and stay up to date
with Walter the Educator™
at WaltertheEducator.com

www.ingramcontent.com/pod-product-compliance
Lightning Source LLC
LaVergne TN
LVHW051921060526
838201LV00060B/4109